How to Sparkle ⊙ Maths Fun

Val Edgar

Brilliant
PUBLICATIONS

We hope you and your class enjoy using this book. Other books in the series include:

Maths titles
How to Sparkle at Counting to 10   978 1 897675 27 4
How to Sparkle at Number Bonds   978 1 897675 34 2
How to Sparkle at Addition and Subtraction to 20   978 1 897675 28 1
How to Sparkle at Beginning Multiplication and Division   978 1 897675 30 4

Science titles
How to Sparkle at Assessing Science   978 1 897675 20 5
How to Sparkle at Science Investigations   978 1 897675 36 6

English titles
How to Sparkle at Alphabet Skills   978 1 897675 17 5
How to Sparkle at Grammar and Punctuation   978 1 897675 19 9
How to Sparkle at Nursery Rhymes   978 1 897675 16 8
How to Sparkle at Phonics   978 1 897675 14 4
How to Sparkle at Prediction Skills   978 1 897675 15 1
How to Sparkle at Word Level Activities   978 1 897675 90 8
How to Sparkle at Writing Stories and Poems   978 1 897675 18 2

Festive title
How to Sparkle at Christmas Time   978 1 897675 62 5

To find out more details on any of our resources, please log onto our website: www.brilliantpublications.co.uk.

Published by Brilliant Publications
Unit 10, Sparrow Hall Farm, Edlesborough, Dunstable, Bedfordshire, LU6 2ES, UK

E-mail: info@brilliantpublications.co.uk
Website: www.brilliantpublications.co.uk

General information enquiries:
Tel: 01525 222292

The name Brilliant Publications and the logo are registered trademarks.

Written by Val Edgar
Illustrated by Chantal Kees

Printed in the UK.
First published in 2001. Reprinted 2009.
10 9 8 7 6 5 4 3 2

© Val Edgar 2001
Printed ISBN: 978 1 897675 86 1
ebook ISBN: 978 0 85747 061 4

Contents

Introduction

How to Sparkle at Maths Fun is a collection of games, practical activities and fun worksheets designed to inspire and reinforce the teaching of maths in the infant classroom.

The book is written to support children working at National Curriculum Key Stage 1 and Scottish National 5–14 Guidelines, levels A and B, and covers a broad range of the work involved in these.

The sheets are ideal for use alongside a programme of classwork, or as single sheets for individual activities. Alternatively they could be grouped for activity booklets for the children to use independently. Several of the sheets, especially the card games and jigsaws, would be best used enlarged and/or photocopied onto card.

The book is based around the theme of jungle animals, with familiar characters leading the children through their learning in three sections.

1. Practical activities (pages 5–21) The sheets in this section are high-interest, hands-on activities involving cut and stick, junk materials, modelling dough etc.

2. Worksheets (pages 22–36) These sheets are work-alone activities which require only coloured pencils as an extra resource.

3. Games (pages 37–46) The children work in pairs or groups for each of these games. They will require dice, counters etc as specified on each page.

For ease of use the **Contents** page provides a breakdown of the main teaching points covered by each sheet. The **Further ideas** page gives some extra ideas for games and whole class activities.

Jungle jigsaw

| 6 six | 7 seven | 8 eight | 9 nine | 10 ten |

| 1 one | 2 two | 3 three | 4 four | 5 five |

Snappy picture line

Cut along the dotted lines.

Use the numbers to put the pictures in order.
You will find a snappy creature.

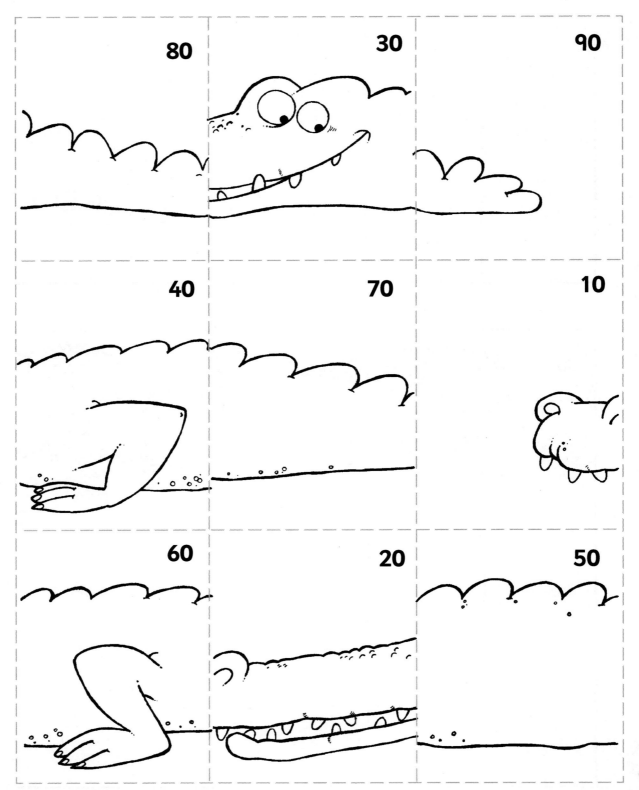

Which animal?

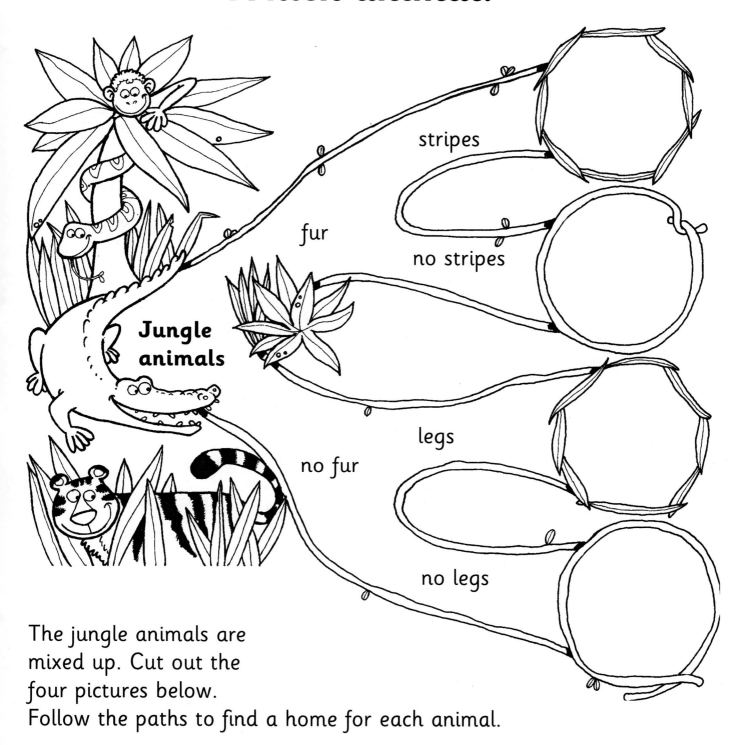

stripes

fur

no stripes

Jungle animals

legs

no fur

no legs

The jungle animals are
mixed up. Cut out the
four pictures below.
Follow the paths to find a home for each animal.

Animal squares

Finish the patterns.

Use the animal squares at the bottom of the page.

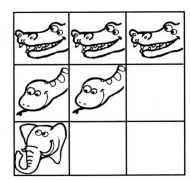

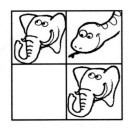

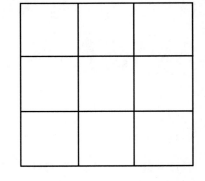

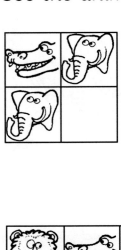

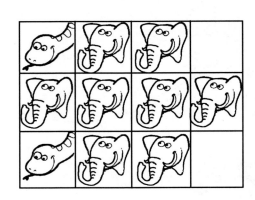

Now make your own patterns.

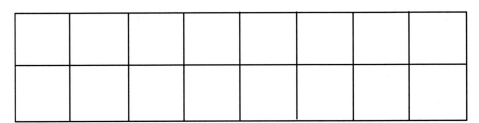

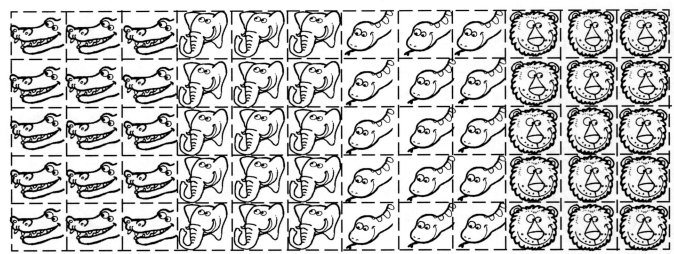

Make it big

Use modelling dough.

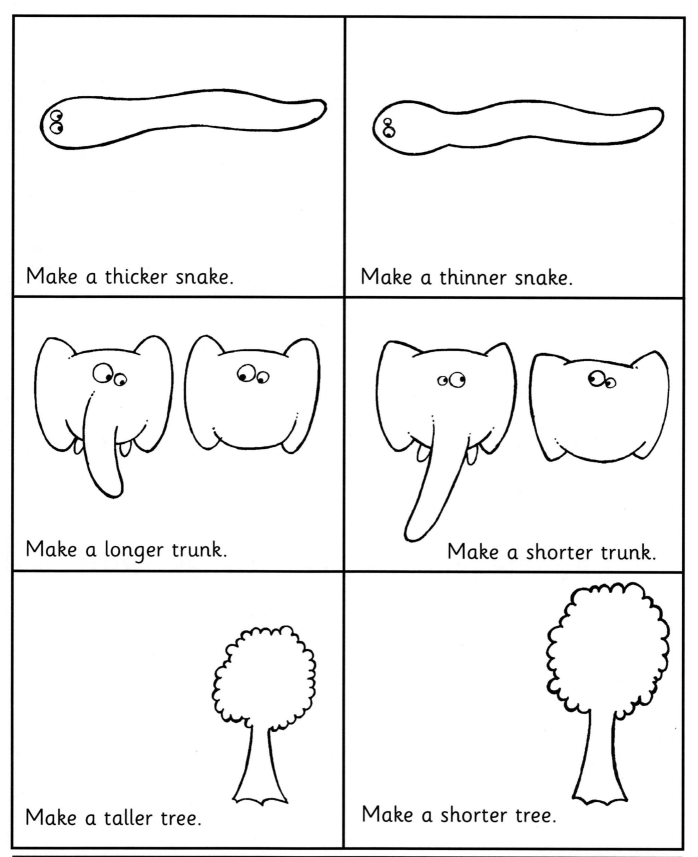

Make a thicker snake.

Make a thinner snake.

Make a longer trunk.

Make a shorter trunk.

Make a taller tree.

Make a shorter tree.

The hundred square

Cut out the animal squares at the bottom of the page.
Stick them in the correct places in the hundred square.

1	2	3	4		6	7	8	9	10
11	12	13	14	15	16	17	18	19	20
21			24	25	26	27	28	29	30
31	32	33	34	35		37	38	39	40
41	42	43			46	47	48	49	50
	52	53	54	55	56	57	58	59	60
61	62	63	64	65	66	67	68	69	70
71	72	73	74	75	76	77	78	79	80
	82	83	84	85	86	87	88	89	
91	92	93	94	95	96		98	99	100

Tallest

Cut out the giraffes.
Put them in order with
the **tallest** first.

Colour the
tallest blue.
Colour the
shortest yellow.

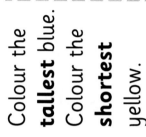

Longest

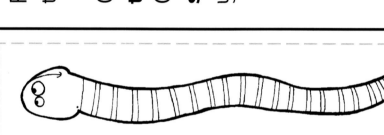

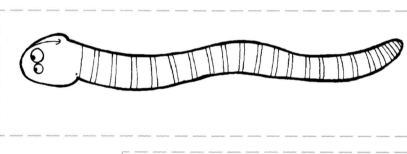

Cut out the snakes.
Put them in order with
the **longest** first.

Colour the **longest**
blue.
Colour the **shortest**
red.

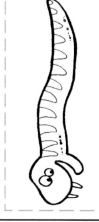

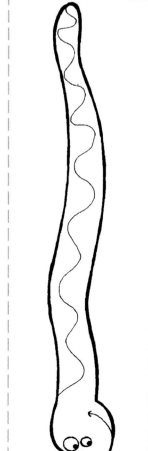

How to Sparkle at Maths Fun

Monkey puzzle

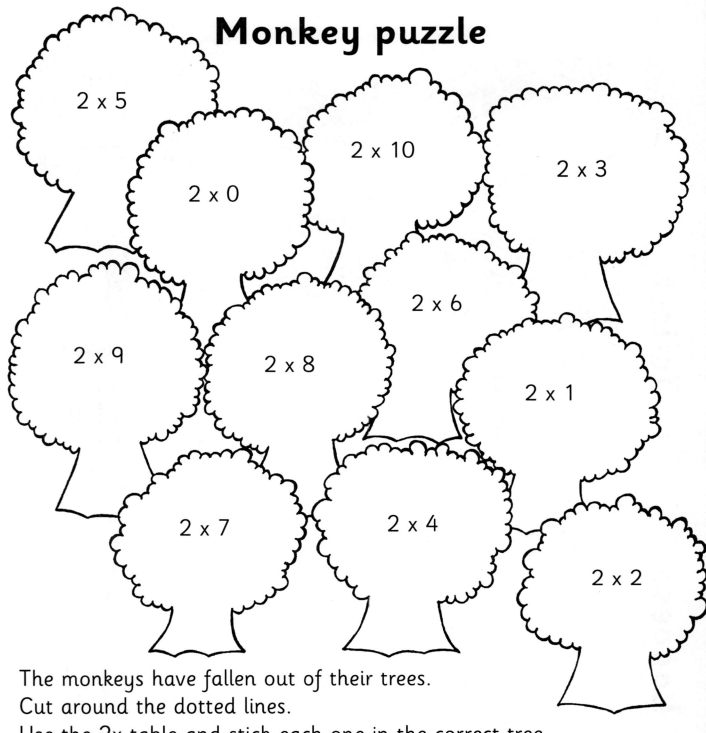

2 x 5

2 x 0

2 x 10

2 x 3

2 x 9

2 x 8

2 x 6

2 x 1

2 x 7

2 x 4

2 x 2

The monkeys have fallen out of their trees.
Cut around the dotted lines.
Use the 2x table and stick each one in the correct tree.

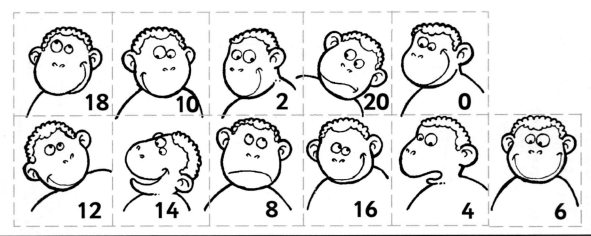

18   10   2   20   0

12   14   8   16   4   6

3D animals

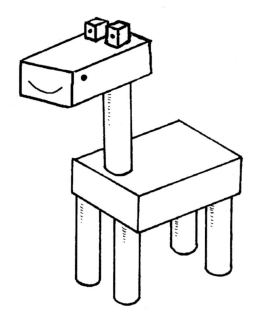

Look at this giraffe.
How many **cubes?** _____
         cuboids? _____
         cylinders? _____

Look at this crocodile.
How many **cones?** _____
         cuboids? _____
         cylinders? _____
         spheres? _____

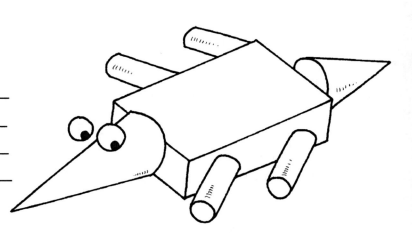

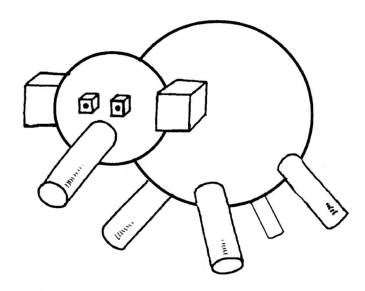

Look at this elephant.
How many **cubes?** _____
         cuboids? _____
         cylinders? _____
         spheres? _____

Make your own animals from 3D shapes.
Ask a friend to count the shapes.

Monkey's busy day

Cut along the dotted lines.
Put the pictures of Monkey in the correct order.
Put the correct clock under each picture.

Snake jigsaw

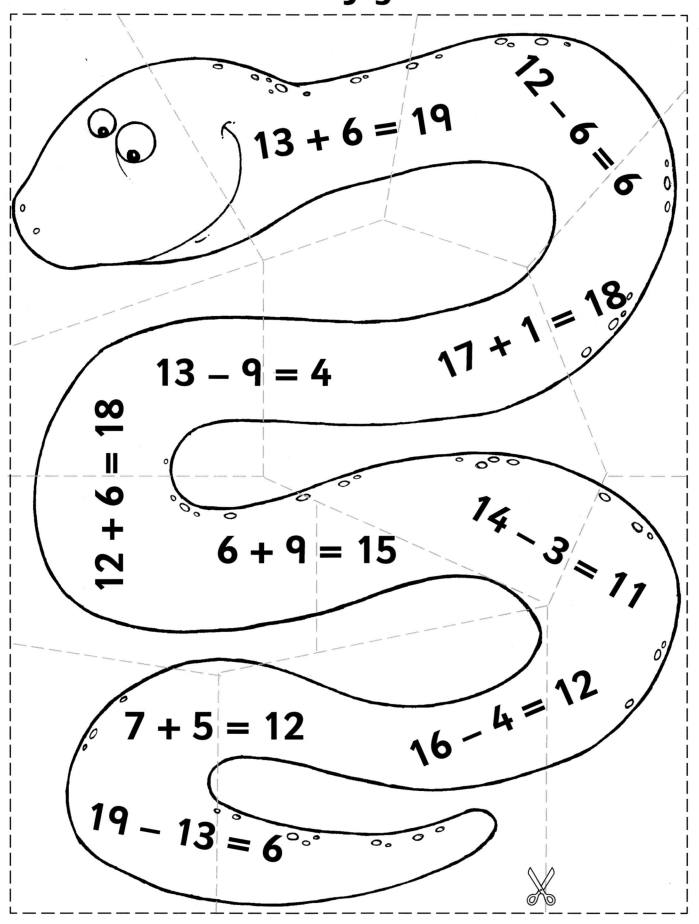

13 + 6 = 19

12 – 6 = 6

13 – 9 = 4

17 + 1 = 18

12 + 6 = 18

6 + 9 = 15

14 – 3 = 11

7 + 5 = 12

16 – 4 = 12

19 – 13 = 6

Zebra and Monkey

Zebra and Monkey are sharing their lunch. Help them to share properly. The food is at the bottom of the page. Use the menus to help you. **Cut and stick the pieces of food onto the plates.**

Monkey's menu
$\frac{1}{2}$ apple
$\frac{1}{4}$ orange
$\frac{1}{2}$ banana

Zebra's menu
$\frac{1}{2}$ banana
$\frac{3}{4}$ sandwich
$\frac{1}{2}$ cake

How much food is left over?

apple_____   banana_____

orange_____   sandwich_____   cake_____

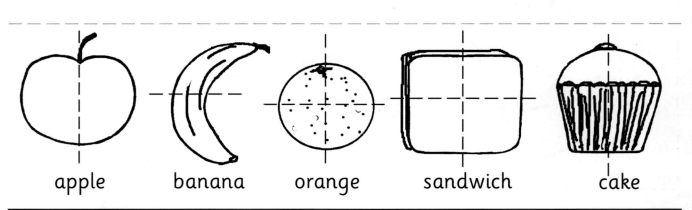

apple   banana   orange   sandwich   cake

Symmetry

Monkey has a mirror and some pictures of his friends. Which pictures have a line of symmetry? **Draw the lines of symmetry using a mirror to help you.**

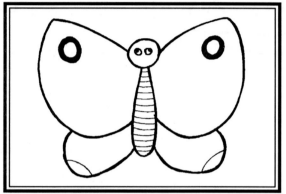

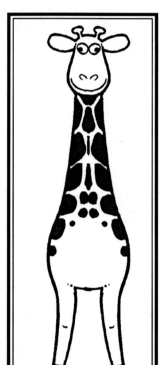

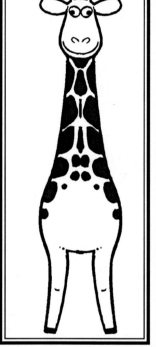

Monkey and me

Monkey used a **cm** tape measure. Here are his measurements.

around head 60 cm

arm 80 cm

around waist 70 cm

foot 36 cm

around wrist 13 cm

height 120 cm

My measurements.

around head _____ cm

around waist _____ cm

arm _____ cm

foot _____ cm

around wrist _____ cm

height _____ cm

Write **Monkey** or **me**.

Who is **taller**? _____

Who is **shorter**? _____

Who has **longer** feet? _____

Who has **shorter** arms? _____

Who has a **thicker** waist? _____

Who has a **thinner** waist? _____

Who has a **wider** head? _____

Butterfly wings

Work out the sums on both sides of each butterfly.

Which butterflies have two wings which match?
Colour them **blue**.

Which do not match?
Colour them **yellow**.

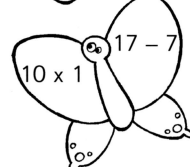

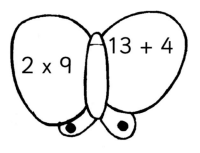

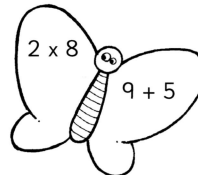

Cut out the sums at the bottom of the page. Match them to these butterflies.

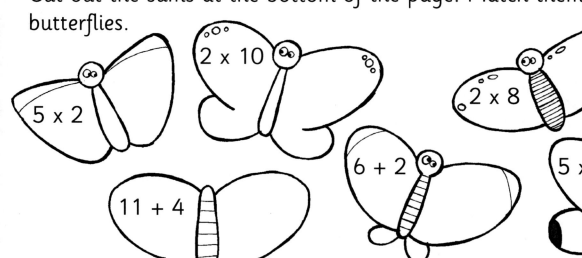

| 19 – 3 | 7 + 3 | 10 x 4 | 4 x 5 | 16 – 8 | 3 x 5 |

Monkey's new toys

Monkey wants some new toys.
Cut and stick the coins for Monkey's toys.

Favourite animals

Which is your favourite jungle animal?

Animal	Number
elephant	
lion	
monkey	
crocodile	
snake	

Ask your group.

Use a tally.

I	=	1
II	=	2
III	=	3
IIII	=	4

Write the number of people who choose each animal in the chart.

Cut out the animal pictures below and stick them on the graph to show what you found.

1. How many people did you ask? _____

2. Which is the most popular animal? _____

3. Which is the least popular animal? _____

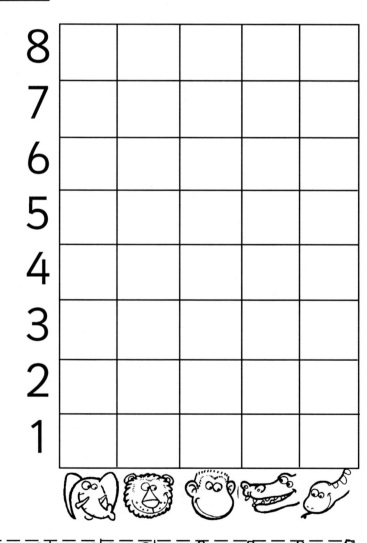

Leopards' spots
Count the spots, match the pairs.

Tick the odd-one-out.

How many?

one	two	three	four	five	six	seven	eight	nine	ten

How many?

giraffes		two	fish		_____
lions		_____	butterflies		_____
trees		_____	crocodiles		_____
snakes		_____	monkeys		_____

Draw **two** more 🦋 butterflies. How many? <u>eight</u>

Draw **one** more 🦁 lion. How many? _____

Draw **two** more 🐟 fish. How many? _____

Draw **one** more 🐛 snake. How many? _____

Dot-to-dot minibeasts

Join the dots together to find the minibeasts.

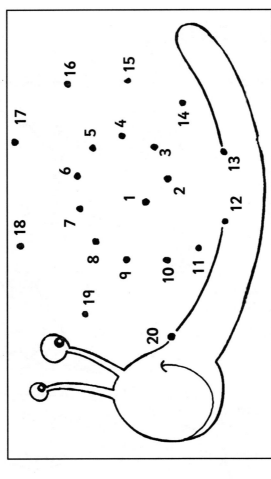

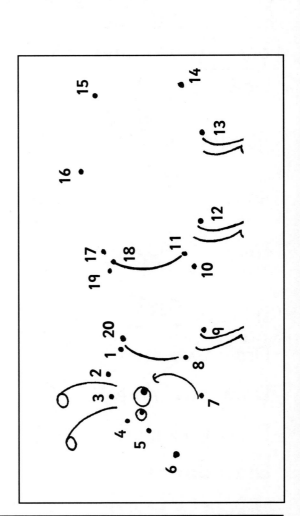

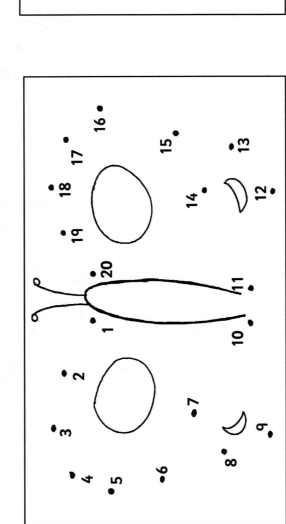

In the jungle

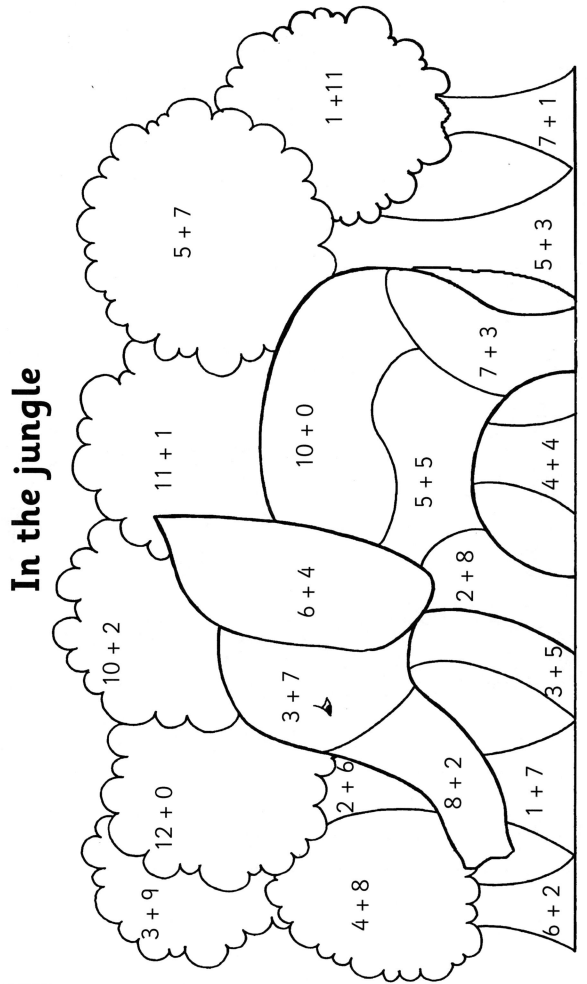

5 + 7

1 + 11

11 + 1

10 + 2

10 + 0

3 + 9

12 + 0

6 + 4

3 + 7

2 + 6

4 + 8

7 + 1

5 + 3

7 + 3

5 + 5

4 + 4

2 + 8

3 + 5

8 + 2

1 + 7

6 + 2

Add the numbers in each shape.
Colour **brown** if the answer is **8**, **grey** if the answer is **10**, **green** if the answer is **12**.

How high?

Monkey has been measuring heights.

How tall is the giraffe? _____ . Is he taller than the tree? _____

Which is the shortest animal?_____

How tall is the elephant? _____.

How much taller is the elephant than the tiger? _____ .

Write down the animals in order from smallest to tallest.

1. _____   2. _____

3. _____

Odds and evens

Colour all the **odd** numbers yellow.

Colour all the **even** numbers blue.

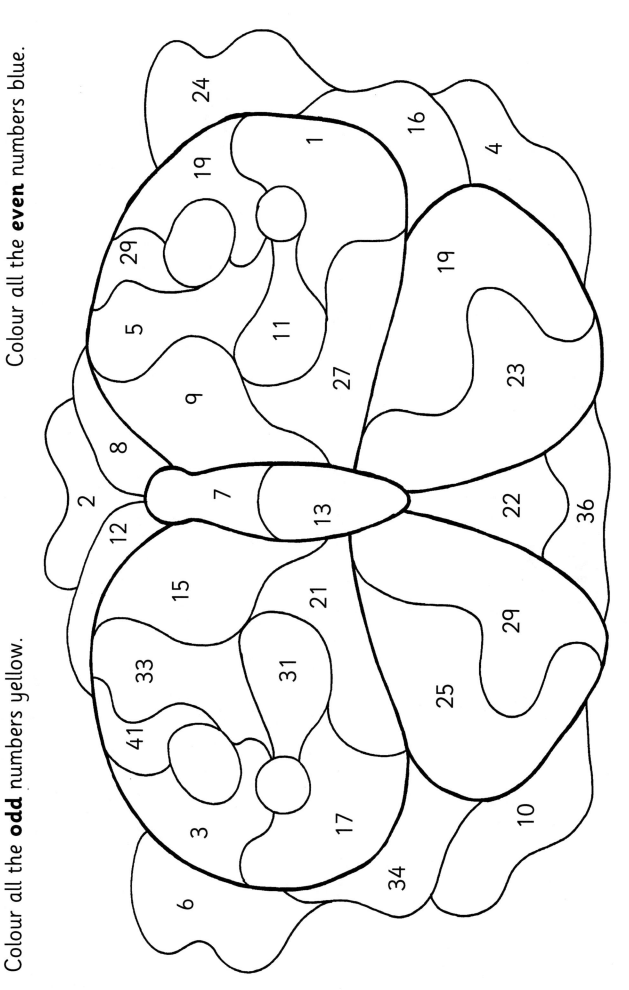

Crocodile 10s

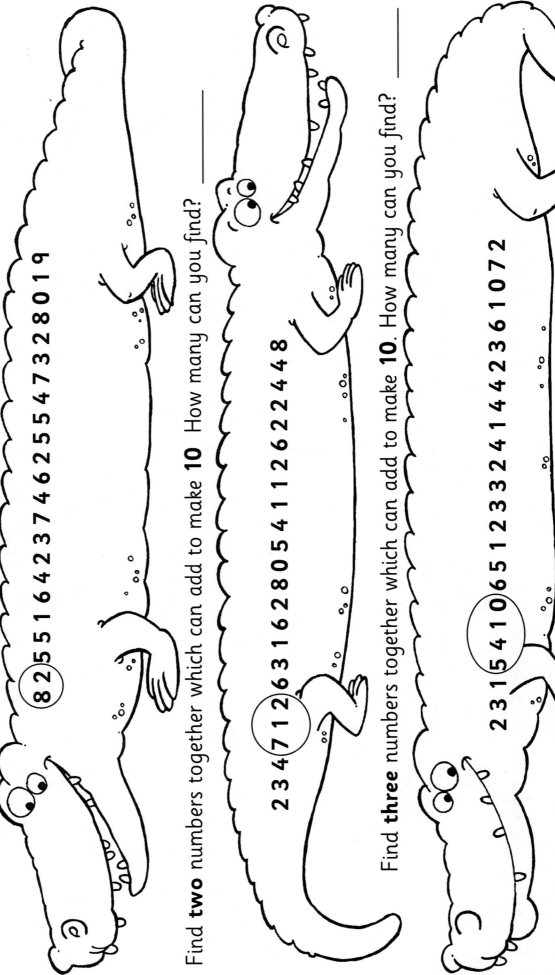

8 2 5 5 1 6 4 2 3 7 4 6 2 5 5 4 7 3 2 8 0 1 9

Find **two** numbers together which can add to make **10**. How many can you find?

2 3 4 7 1 2 6 3 1 6 2 8 0 5 4 1 1 2 6 2 2 4 4 8

Find **three** numbers together which can add to make **10**. How many can you find?

2 3 1 5 4 1 0 6 5 1 2 3 3 2 4 1 4 4 2 3 6 1 0 7 2

Find **four** numbers together which can add to make **10**. How many can you find?

Snake patterns

Colour the snakes. Finish the patterns.

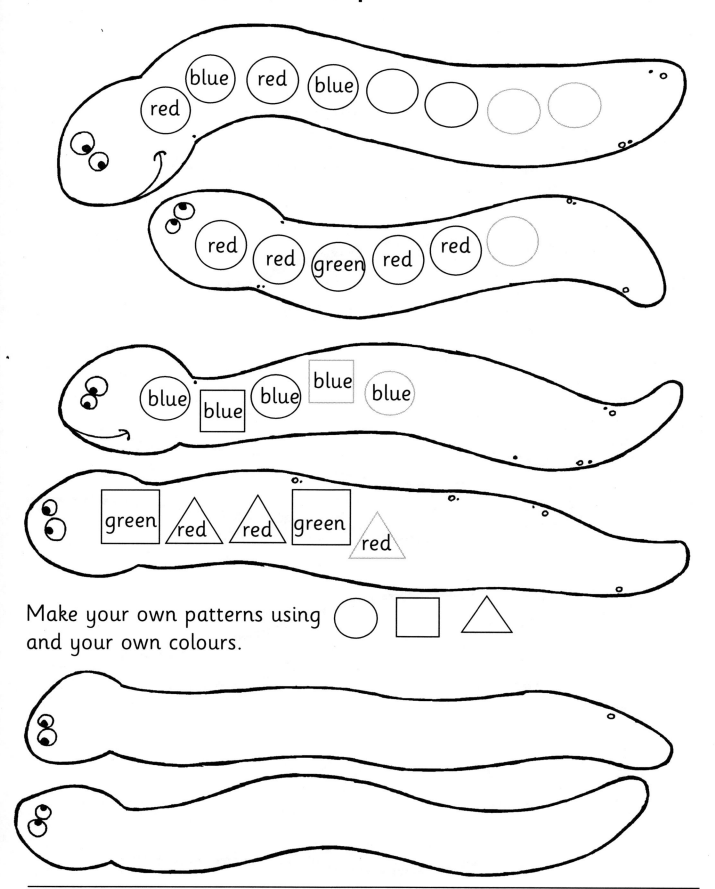

Make your own patterns using ◯ ▢ △
and your own colours.

Giraffes' necks

Match to make **20.**

Match to make **18.**

Match to make **15.**

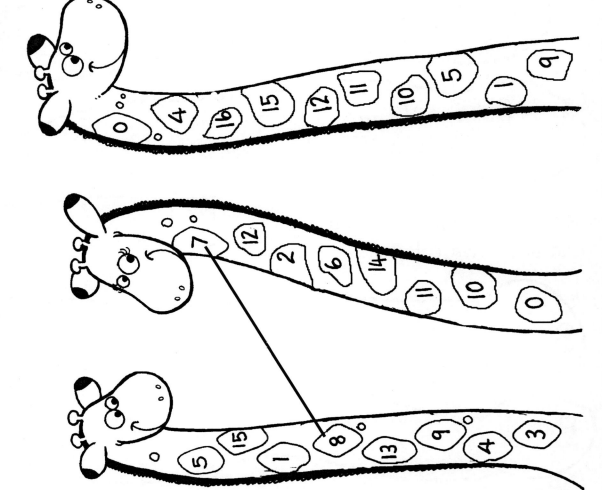

The banana trail

One of the monkeys has been eating the bananas. Colour all the 5x table leaves to find a trail from the bananas to the greedy monkey.

Hungry faces

Follow the instructions to find the animals with the hungry faces.

Start at 2 and count in 3s.

Start at 2 and count in 4s.

Start at 2 and count in 2s.

Start at 5 and count in 2s.

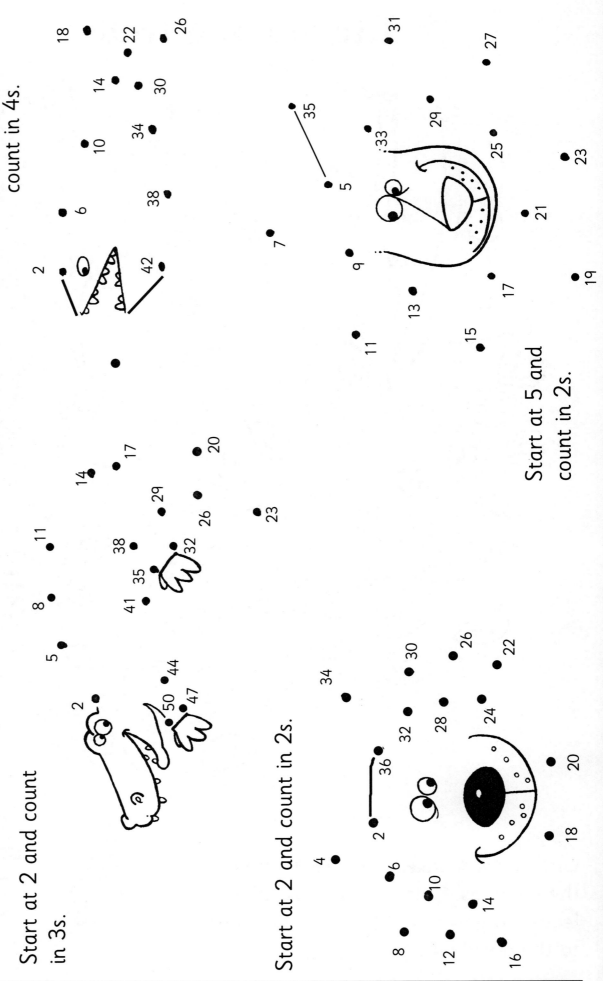

Hidden numbers

Find the numbers hidden in the elephant's back.

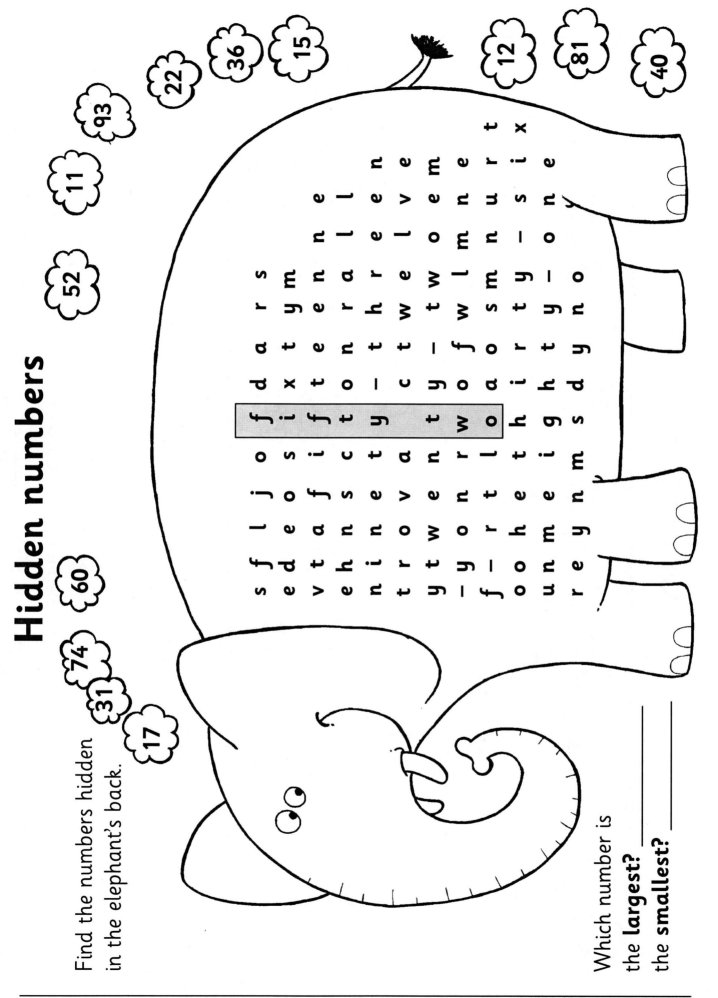

Which number is

the **largest?** _____

the **smallest?** _____

Groups of 2

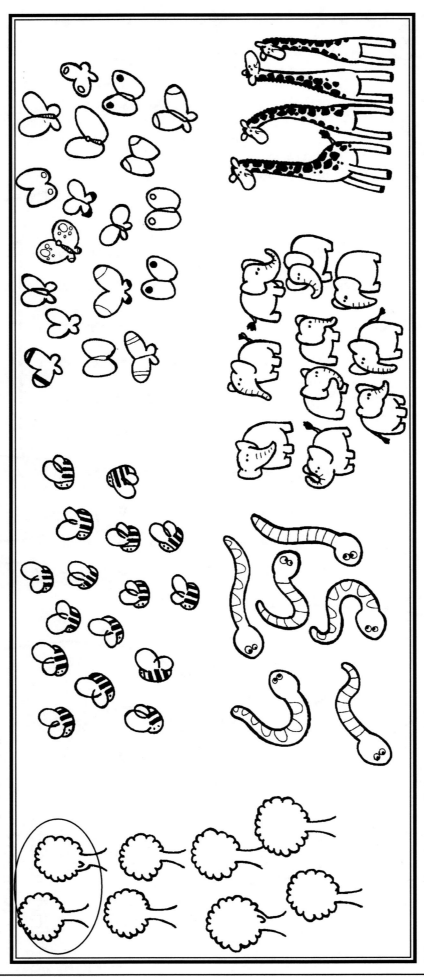

How many **trees?** _____
Put the trees in groups of 2.
How many groups? _____

How many **snakes?** _____
Put the snakes in groups of 2.
How many groups? _____

How many **bees?** _____
Put the bees in groups of 2.
How many groups? _____

How many **elephants?** _____
Put the elephants in groups of 2.
How many groups? _____

How many **butterflies?** _____
Put the butterflies in groups of 2.
How many groups? _____

How many **giraffes?** _____
Put the giraffes in groups of 2.
How many groups? _____

Caterpillars

Which caterpillar ate a hole in which leaf? Use the **10x** table to match.

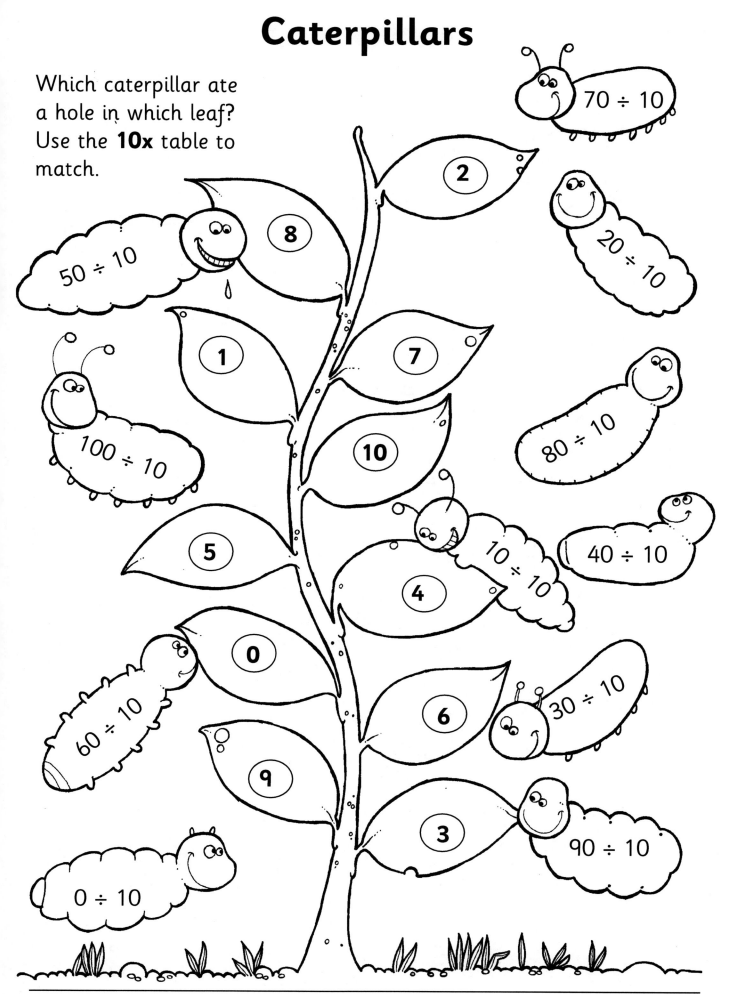

Animal patterns

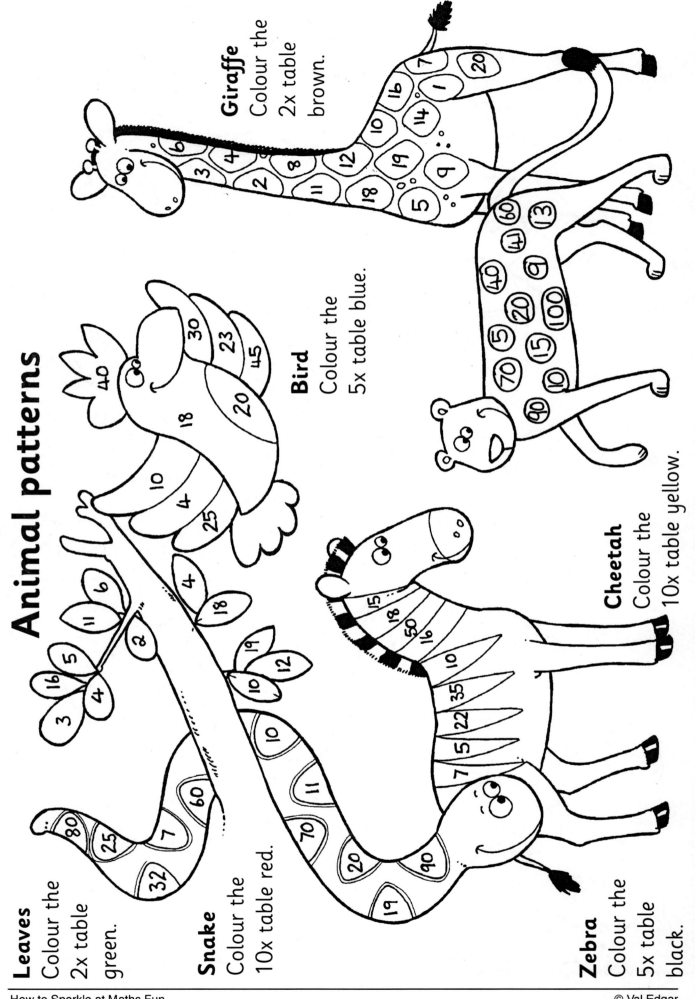

Giraffe
Colour the 2x table brown.

Bird
Colour the 5x table blue.

Cheetah
Colour the 10x table yellow.

Leaves
Colour the 2x table green.

Snake
Colour the 10x table red.

Zebra
Colour the 5x table black.

Money pairs

Cut out the cards. Use them to play **Pairs**.
Shuffle the cards and place them face down. Players take turns to turn over two cards to try to find a matching pair. Who can find most pairs?

Picture race

Play this game with a partner. Choose a picture. Take turns to roll a die. Who can colour their picture first?

colour a **square**

colour a **triangle**

colour a **circle**

colour a **rectangle**

miss a turn

miss a turn

Snap

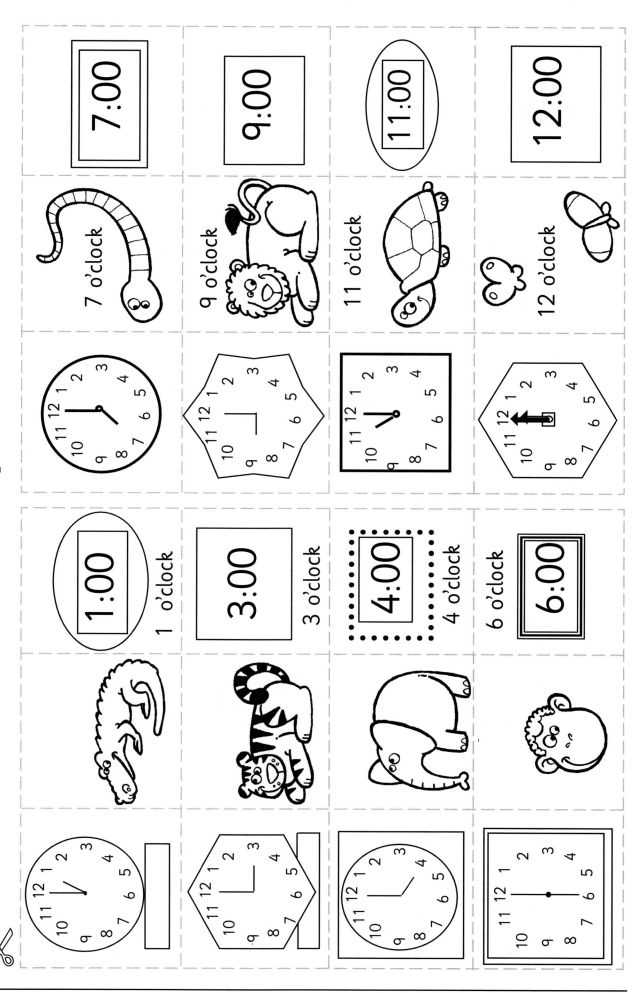

7:00

9:00

11:00

12:00

7 o'clock

9 o'clock

11 o'clock

12 o'clock

1:00

3:00

4:00

6:00

1 o'clock

3 o'clock

4 o'clock

6 o'clock

Jungle flowers

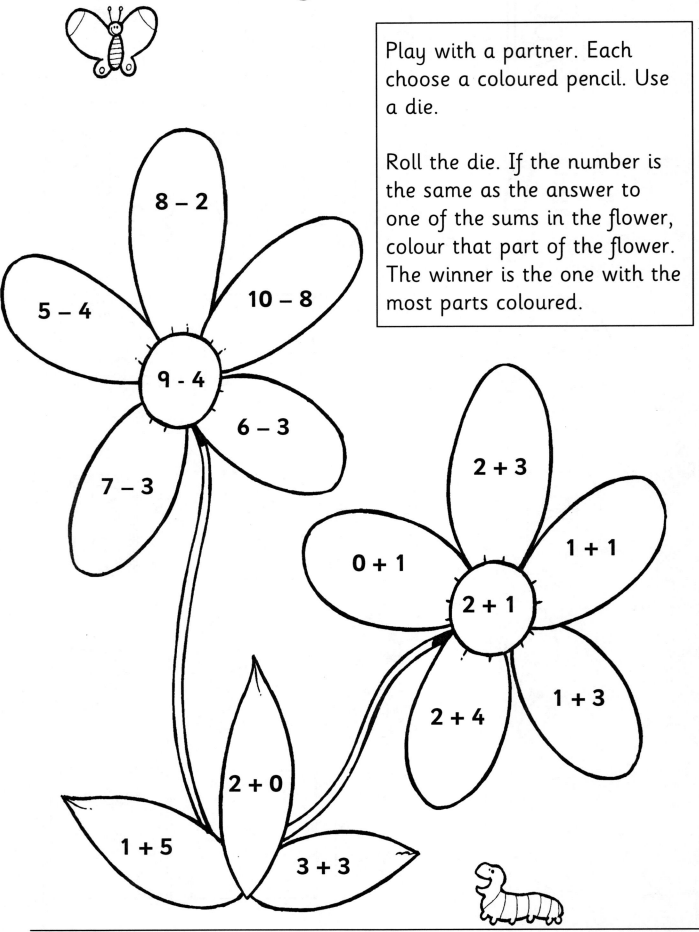

Play with a partner. Each choose a coloured pencil. Use a die.

Roll the die. If the number is the same as the answer to one of the sums in the flower, colour that part of the flower. The winner is the one with the most parts coloured.

8 – 2

5 – 4

10 – 8

9 - 4

6 – 3

7 – 3

2 + 3

0 + 1

1 + 1

2 + 1

2 + 4

1 + 3

2 + 0

1 + 5

3 + 3

Odd spots

Play with a partner. Choose one giraffe each. Use two dice.

Roll the dice and add the numbers together. If the total is an **odd number** colour a patch on your giraffe. Who can colour all of their giraffe's patches first?

July in the jungle

You will need a die and a counter for each player.

Roll the die and move through the calendar page. Who can reach Friday 31st first?

Sun	Mon	Tue	Wed	Thur	Fri	Sat
			1 Start	2	3 Party! Take another turn	4
5	6 Dentist! Miss a turn	7	8 Birthday! Go to next Wednesday	9	10	11 Sick! Go back one week
12	13	14 Accident! Go back 7 days	15	16 Holiday! Move to 19th	17	18
19	20 Party! Jump forward one week	21	22	23	24 Birthday! Go forward two days	25
26	27	28	29 Go back to 3 days	30	31 Finish	

Take away snake

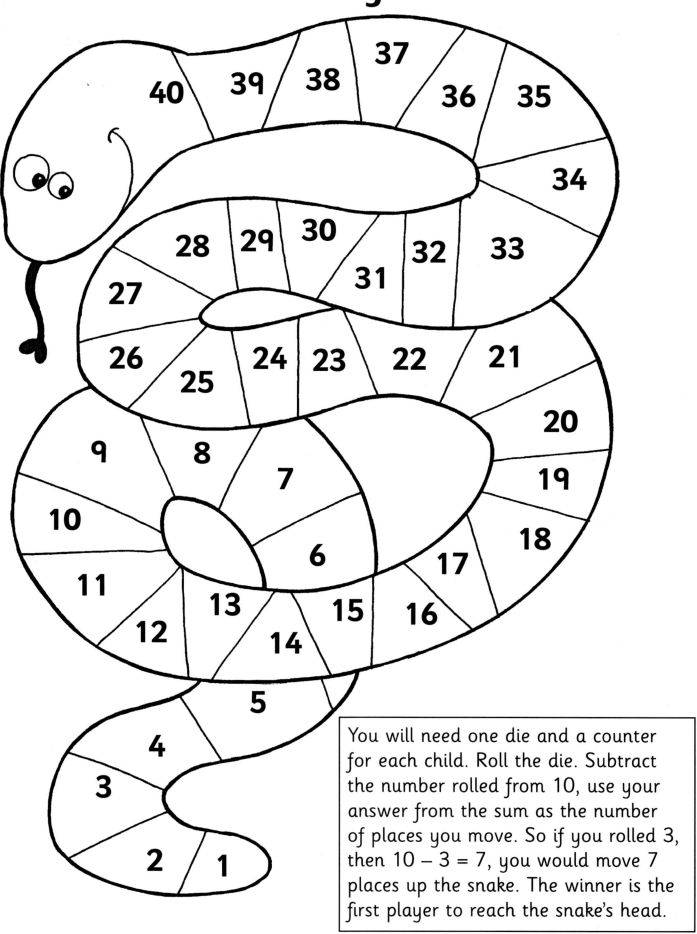

You will need one die and a counter for each child. Roll the die. Subtract the number rolled from 10, use your answer from the sum as the number of places you move. So if you rolled 3, then 10 − 3 = 7, you would move 7 places up the snake. The winner is the first player to reach the snake's head.

Shapes bingo

Teacher's note:
Cut out the four bingo cards.
Cut out and make the die.

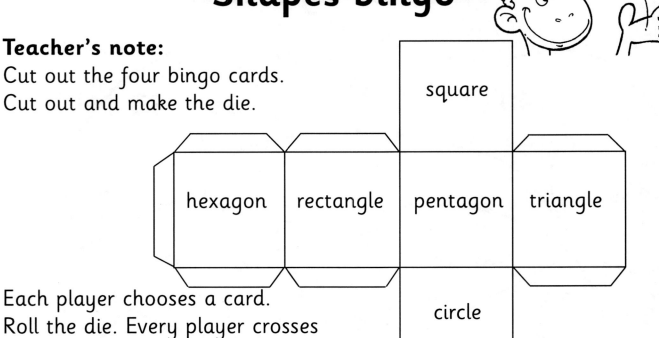

Each player chooses a card.
Roll the die. Every player crosses
off that shape **once** on their card.
Roll again. Continue until one
player has crossed four shapes in
a row, and shouts **Bingo**!

Climb the vine

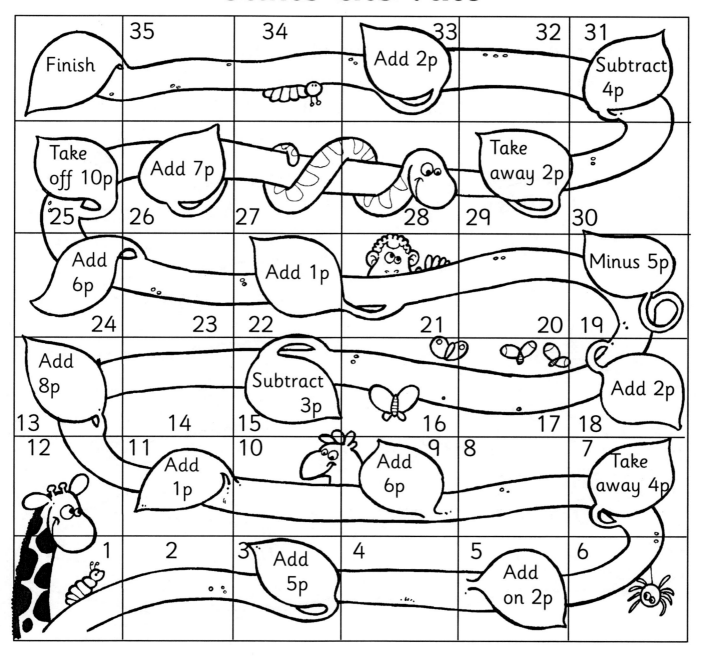

You will need a die, counters, pencils and paper.
Write down 20p for each player like this:

This is your starting money.
Each time you land on a money leaf add
or subtract the amount on the leaf. Write
down your new total like this:

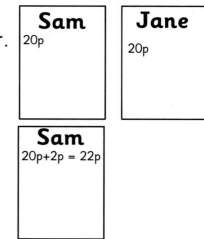

Sam	Jane
20p	20p

Sam
20p+2p = 22p

When everyone has reached the finish, who has the most money?

Crocodile capers

Roll a die and move in the direction of the arrows.
If you land on a number in a circle, divide the number by **5**. Tell the other players your answer. If you are correct, move on 3 spaces.

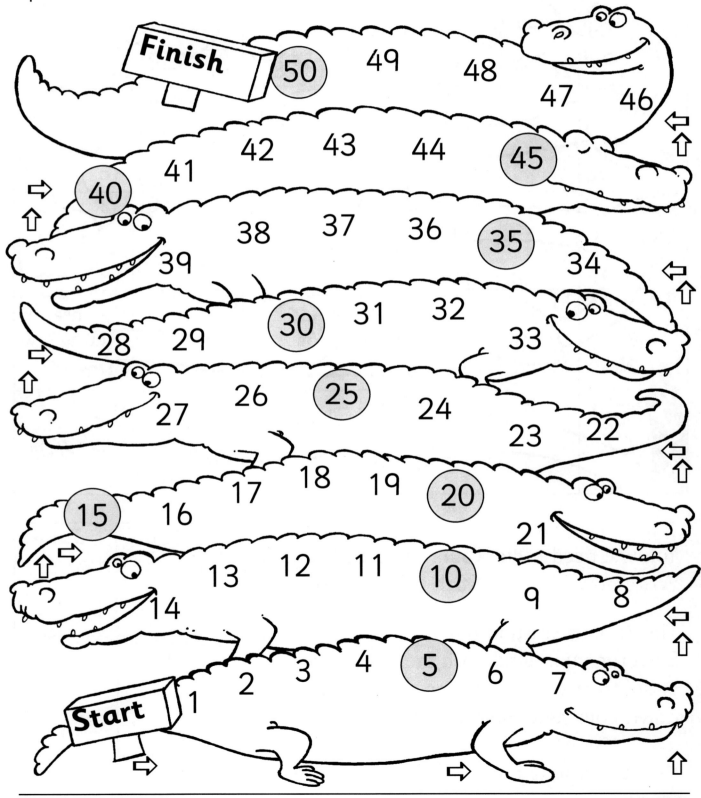

Further ideas

Jungle jigsaw – page 5
Cut the sheet into strips and use as a reference aid until children have grasped the numerals and words.

Which animal? – page 7
As a class, attempt to classify other jungle animals by distinguishing features.

Make it big – page 9
Game in pairs. One child makes an animal and challenges the other to make one with a bigger nose/shorter tail/ thicker legs.

The hundred square – page 10
Game in pairs. One child covers five squares with counters while the other looks away. Second child then guesses the numbers underneath.

Longest, tallest – page 11
Ask children to bring teddies from home and sort them in different ways.

Monkey puzzle – page 12
Use the sheet to play tiddlywinks. Flick a counter onto the sheet and if it lands on a tree give the answer to the sum. Each correct answer scores one point.

3D animals – page 13
Create a class display of 3D animals made by the children; encourage discussion of shape properties.

Zebra and Monkey – page 16
Have a picnic lunch in class to demonstrate halving and quartering fruit, cakes and sandwiches.

Symmetry – page 17
Make symmetrical butterflies, snakes and even tigers by painting, folding and pressing paper.

Monkey and me – page 18
Take this idea one stage further by comparing results within the group, looking for tall**est**, long**est** etc.

Butterfly wings – page 19
A class display of similar wings is very useful for reinforcement of number bonds and multiplication facts.

Monkey's new toys – page 20
Challenge children to find change from £1/100p for each item, and to find the correct coins for the change.

Favourite animals – page 21
Create a class graph of favourites, and discuss.

Money pairs – page 37
These cards can be copied twice and used to play Snap.

Crocodile capers – page 46
Use the game to reinforce other teaching points; eg every time you land on a circle double the number or add 20.

Answers

3D animals – page 13
giraffe – 2, 2, 5
crocodile – 2, 1, 4, 2
elephant – 2, 2, 5, 2

Zebra and Monkey – page 16
$^1/_2$ apple, 0 banana,
$^3/_4$ orange, $^1/_4$ sandwich,
$^1/_2$ cake

Symmetry – page 17
symmetrical – butterfly, tiger,
tortoise, bee, elephant

Butterfly wings – page 19
Matching – 2x2, 9–5; 20–5,
3x5; 9–5,2x2; 10x1, 17–7;
Not matching – 6+3, 5x2;
2x9, 13+4; 2x8, 9+5;

5x2, 7+3; 2x10, 4x5;
2x8, 19–3; 6+2, 16–8;
5x8, 10x4; 11+4, 3x5

Monkey's new toys – page 20
15p – 10p, 5p
82p – 50p, 20p, 10p, 2p
63p – 50p, 10p, 2p, 1p
27p – 20p, 5p, 2p
(other variations possible)

Leopards' spots – page 22
odd–one–out – 8 spots

How many? – page 23
two giraffes   seven fish
three lions   six butterflies
three trees   one crocodile
four snakes   two monkeys

eight butterflies
four lions
nine fish
five snakes

How high? – page 26
4m
no
tiger
3m
2m
tiger, elephant, giraffe

Crocodile 10s – page 28
(8,2) (5,5) (6,4) (3,7) (4,6) (5,5) (7,3)
(2,8) (1,9); nine.
(7,1,2) (7,1,2) (6,3,1) (2,8,0) (5,4,1)
(2,6,2) (2,4,4); six.
(5,4,1,0) (2,3,3,2) (3,6,1,0); three.

Hidden numbers – page 33

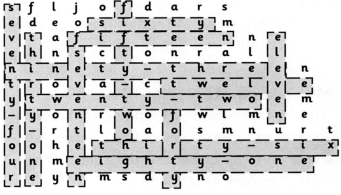

largest - ninety-three
smallest - eleven

Groups of two – page 34
trees – 8, 4   bees – 16, 8
butterflies – 18, 9   snakes – 6, 3
elephants – 10, 5   giraffes – 4, 2

Lightning Source UK Ltd.
Milton Keynes UK
174491UK00001B/15/P